Vedic Math Tips

John Carlin

Copyright © 2014 John Carlin

All rights reserved.

ISBN:1501068660
ISBN-13:978-1501068669

CONTENTS

1	Squaring Two Digits in Two Steps	1
2	Squaring Three Digits in Two Steps	9
3	Use Aliquot Parts	15
4	Three Digits any Numbers	25
5	Solve for Something Else	33

1. Squaring Two Digits in Two Steps

You can square any two digit number by taking the sum of the units column and then carrying forward that amount to the larger tens' digit. Then you multiply that amount by the smaller tens' digit. That result is equal to the first vertical calculation and the sum of your cross products. The final step would be to the last vertical calculation. It sounds more complicated than it actually is.

Almost everyone knows that you can square a number that ends in five by multiplying the tens' digit by one unit more than its value and tacking on 25. For example, if you wanted to square 25, you could multiply 2 x 3 to get 6. The next step would be to tack on 25. Your final result would be 625. In a similar fashion, 35 squared would be three times four equals 12, and then you tack on 25 to get 1225. You could do the same calculation for any two digit number that ends in five, such as for the numbers 15, 25, 35, 45, 55, 65, 75, 85, or 95.

If you think about it, the calculation that we are doing is exactly what we described in the very first paragraph above. Since the total of the units' digits is 10, we are carrying forward 10 to the tens' column. That then becomes a one to be added to one of the two numbers. You could do the same thing no matter what the units add to. If you wanted to square 11, you could add the units' digits to get 02. Now carry that forward to the tens' column, and add to one of the tens. One would become 1.2x10. So now we are working with 12 and we are multiplying it by the remaining one in the tens' column. That would give us our first vertical calculation and the sum of the cross products all in one step. All that remains now is to link up the product of the units' column which is one times one. Our final result becomes 121. We just squared 11 in two steps.

Let's try another example. Let's square 24. The units add up to eight. So we are taking 28 and doubling it to get 56. We know that 4x4=16. If we link 56/16, we will have our final answer of 576. I hope that you are brimming with ideas about squaring numbers at this point. Some numbers in particular would be especially adaptive to this method. Anything that began with a one, for example, would be absolutely easy. Once you added the carry forward amount, you will have your cross product right under your nose. Let's square 17. 7+7 equals 14. Add ten to it to get 24. To that, we have to link 7x7, which is 49. Our final link would be 24/49= 289.

In a similar fashion, 19 squared would be 28/81= 361. Squaring anything in the 20s would also be particularly easy. We are just doubling the sum of our carry forward amount and one of the tens' digits. For example, 28 squared would be 36x2=72. To this, we would link 8x8= 64. Our final result would be 72/64=784. 29 squared would become 76/81= 841. Numbers in the 30s, 40s, and 50s would all be particularly easy.

In that Two digit squares are so useful, I generally recommend that you just memorize them if you really want to become a mental math whiz. This method would be a close second to rote memorization. In one of my earlier books, I had a long section about the importance of being able to multiply, double and triple digit numbers by a single digit. I hope that this demonstration of carrying forward and doing double-digit by single digit multiplication shows you the power of these two items. The method extends a shortcut that everyone knows to all double digit numbers that you would want to square. You don't need no stinking calculator to square a two digit number!

One often overlooked fact is that all of these methods are bidirectional. For example, let's look at the example of 25 squared again. We could reverse the way that we look at the columns and proceed from there. In this case then, you add the sum of the tens' column and carry it backwards. That would give you the number 45 in the units' column. Then you would multiply that by the remaining five in the units' column. That would give you 225. To that you would add the product of the tens' column which is 2x2 equals four. The four would go on the front end of the 225, in the form of 04/225= 625.

In a similar fashion, you could square 35. You would add the threes together to get six. Bring it over to the units' column to get 65. Multiply that by five to get 325. Then link it with 3x3. Your final result would be 09/325= 1225.

I hope that you realize that this is a bit like Tesla and Edison arguing over AC and DC current. AC current is, of course, bidirectional. DC current is one directional. The big argument back then was over which was better. Realizing the advantages of bi-directionality is sort of like realizing the advantages of AC current. You can transmit your leverage or power so much farther by using bi-directionality. We mentioned how effective this technique is for squaring numbers in the 10 through 50 category. You can double that effectiveness now through bi-directionality to apply to squares that end in the numbers one through five.

For example, if you want to square 92, you might want to consider adding the tens' columns and carrying that sum backwards to one of the twos. That would give you 182. Now just double that number to get 364. To that link the product of 9x9. The final result would be 81/364= 8464. That might be easier than multiplying 9x94=846. With bi-directionality. you have an option to work the smartest way. You don't have an option to work smart using the grade school method you have been stuck with the majority of your math life. The option to work smart is huge. In the three digit world, you have options about how to split the number, and now even an option that allows you to work after the split with the easiest numbers. How great is that?

The best way to use bi-directionality is to look at the number that you're squaring, and choose the column with the smaller digit in it as the column that you will use to multiply by. The column with the larger number would always be the column that you sum and carry either backward or forward to. In this way, you would always have to multiply by the smallest number possible. It's easier to add the larger numbers, and multiply by the smaller number than vice versa. For example, if you wanted to square 19, you would carry forward. Conversely, if you wanted to square 91, you would add the nines, and carry backwards.

Exercises:

Square these numbers mentally, and just write down your answers. Work them according to number order so that you are taking advantage of bi-directionality and get to experience it on real problems.

1. 13
2. 14
3. 22
4. 23
5. 32
6. 36
7. 41
8. 49
9. 38
10. 52
11. 57
12. 63
13. 69
14. 74
15. 78
16. 82
17. 86
18. 92

19. 97
20. 99

Answers:

1. 169
2. 196
3. 484
4. 529
5. 1024
6. 1296
7. 1681
8. 2401
9. 1444
10. 2704
11. 3249
12. 3969
13. 4761
14. 5476
15. 6084
16. 6724
17. 7396
18. 8464
19. 9409
20. 9801

2. Squaring Three Digits in Two Steps

You can square any three digit number in a similar fashion to how we did it for two digit numbers. This time, you split the three digits into a one digit number and a two digit number. The choice of putting the two digits first or last is entirely up to you. An example might be the best way to give you some ideas about how you might decide to split a three digit number that you wish to square. If you wanted to square 124, you could split it into 1/24, or 12/4. Which would you choose? Any number that starts or ends with 1 is pretty simple to square just using the base and overage method. I know the answer to this one immediately. You are 24 units over a base of 100, therefore add 24 to 124 to get 148. Now link that to the square of the amount of the overage. 24 squared is 576. Our final answer is 148/576=15376.

The generally accepted easiest way to do this one without the base and overage narrative is exactly the same. Split 124 into 1/24. The units add up to 48, so we carry that forward to the 100. Our new number to be multiplied by 1 is 148. 148x1 is 148. That is the first part of the solution. I know my two digit squares, so the rest is quite easy,. Just link 148 and 576 to get 15376. The other approach might be to split this into 12/4. I know that 12x12=144, so I can proceed to use vertical and crosswise. From a fractions standpoint ,12/4= 3. If I double that, I know that my middle component is 6 times 4x4. The links would involve putting together 144/96/16= 15376. I can also carry back the sum of 12+12. Then multiply 244x4 to get 976. This would result in assembling 144/976=15376. We just solved this five different ways. We used vertical and crosswise, fractions, carry forward, carry backward, and over a base. They all worked, but the easiest, for me at least, was to treat this as an over a base of 100 problem, or a carry forward with a split of 1/24, so that I could use the power of the identity property and just link two components 148 and 576. The important point is that just seeing one as the first digit immediately tells you the easiest way to solve this problem.

Another example might involve just turning the number around and squaring it. Let's look at squaring 421. In this case, you could split the number into 4/21, or 42/1. I know that 42x42=1764, or at the very least, I could get that part by using a fractional approach. 4/2x2=4, so my components to link are 16/16/4=1764. Now I just link 1764/841= 177241. Again, once you see the one as a digit, that is the number that you want to gravitate towards. The base narrative yields the exact same result. 421 is 420 units over a base of 1. Add that overage to 421 to get 841. Now link that with the product of 42x42.

I always split my three digit squares so that the single digit is the smaller of the lead or trailing digit. In other words, I split it so that my two digit square is as big as I can get it. It is consistent method, so I don't lose a lot of time kicking the tires so to speak. The procedure should be to carry back or forward towards the smallest number of the two digits.

I hope that this gives you a feel for

squaring three digit numbers. Every multiplication is different, but at least you have a guideline for doing the work. The way you really get good at this is to work with the numbers. Remember the 10,000 hour rule. It takes a long time to get really expert at anything. You can get relatively proficient very rapidly though. That part of the journey will be extremely rewarding, and may be the most fun.

Exercises:

Square these three digit numbers. Show how you would split them, the two step components, and the final answer.

1. 789
2. 456
3. 123
4. 741
5. 987
6. 654
7. 321
8. 147
9. 758
10. 367
11. 532
12. 235
13. 498
14. 894
15. 265

Solutions:

split into:	Components:	answers:
1. 7/89	6146/7921	622521
2. 4/56	2048/3136	207936
3. 1/23	146/529	15129
4. 74/1	5476/1481	549081
5. 98/7	9604/13769	974169
6. 65/4	4225/5216	427716
7. 32/1	1024/641	103041
8. 1/47	194/2209	21609
9. 7/58	5712/3364	574564
10. 3/67	1302/4489	134689
11. 53/2	2809/2124	283024
12. 2/35	540/1225	55225
13. 4/98	2384/9604	248004
14. 89/4	7921/7136	799236
15. 2/65	660/4225	70225

3. Use Aliquot Parts

Just as you can square numbers with the sum and difference method, you can also multiply numbers that are not identical. Let's look at a commonly used example of mine. We want to multiply 39x21. We know that since the units' column totals to 10, and there is a difference of one between the two digits of the tens' column, that the cross product is in fact 21. No multiplication is necessary. From there, you can quickly figure out that 39x21 = 819. If we didn't know all that, we could still derive the answer by noting that the units' column sums up to 10. Then we would carry forward the 10 to the tens' column, where it would become 01 to be added to the larger number three. Then we would multiply (3+1)x2= 8. Now we have the first digit of the solution. For the second digit, we could note that the difference between the original numbers in the tens column is one. Now we multiply that one by the one in 21. Or to put it another way, we multiply that difference of one by the units' digits of the smaller number. In this case, one times one equals one. That number is our second digit, We are two thirds of the way to our answer. The final digit is merely 9x1, or the vertical product units' digits. This same approach works for all two digit calculations.

If we wanted to multiply 78x32, we could do it by noting that the 2 and 8 are tens' complements. Now we add one to the 7 in 78, and multiply that (7+1) by 3 to get 24. The second step is to find the difference between seven and three, which is four. Then we multiply that 4 by the 2 in 32. That gives us 08. Our final component is 8x2= 16. Let's put those components together; we have 24/08/16= 2496.

In fact, regardless of what the units' digits add up to, we can still use the sum and difference method to get an answer. If we wanted to multiply 79x33, we could do it using the same method. Since 9+3 equals 12,

we will now carry forward to the seven in 79.
That gives us 82. Then we take 82, and
multiply that by 3 to get 246. The difference
between the tens' digits is still 7-3= 4, and we
multiply that difference by the 3 in the units'
column of 33. Our second component is 12.
The final component is 27. Let's link them
together to get 246/12/27= 2607. This may
not seem like much of an improvement over
the standard vertical and crosswise method.
I find that it really does help the calculation
flow, however, because carrying forward
automatically links some of the digits
together. You don't do a calculation, set it
down, do another one, and then set it down.
The calculation melds together a little better
using this approach. The hardest part of
mental math once you get the hang of it is
keeping track of the digits and components
as you calculate. Anything that facilitates
that is extremely useful.

    I'll show you another reason why this
method has some advantages. Let's change
the example we just did slightly. Let's say
that we want to calculate the product of
79x35. Now the units' digits add up to 14,
and we carry just the 10 part of the 14 over
to the 7 in 79, 7+1= 8 which we need to

multiply by 3. That calculation gives us 24. Now we take the difference between the seven and the three which is four. Note that we also have a difference of four from when we left 4 on the units' side earlier. Our cross product is going to be 4 times the sum of the digits in 35. That means our cross product is 4x(3+5)= 32. Our components become 24/32/45= 2765.

Using this little trick with the difference equaling the sum works in all kinds of situations that you don't normally recognize. For example, if you wanted to multiply 125x7, you could easily do it by recognizing that 5+7=12, and 12-0 also equals 12. From there, just link 12x7 with 5x7. The components would be 84/35= 875. I don't even think of this process as multiplication, but in fact it is.

If we expand this process to a 3 digit by 2 digit multiplication, an example might be 199x19. This has lots of nines in it, and appears to be a little onerous to do. Let's see what we can do with it. The nines add up to 18, and the difference between 19 and 1 is also 18. Our components become 19/180/81= 3781. That middle component is 18x(9+1)=180.

There are any number of examples where this works. We can use it on mixed numbers of digits, or on problems where the numbers of digits are the same. In some cases where the digits are the same we have the added step of taking the extra digit over to the tens' column. In situations where the units' sum of the digits is less than ten, we can dive right in and use the method straight away. If we multiply 98x29, we would make the 9 a 10 since 8+9=17. Our first component would be 20. After that, the 7 in 17 and the difference of 7 in 9-2 tells us that the second component would be 7x(2+9)=77. The final result would be 20/77/72= 2842. If we wanted to get 96x21, we could just write down the components as 18/21/06=2016. The calculation couldn't be much easier.

The sum and difference method has some of the features of the open carry method that we discussed in a previous book. For example, if we wanted to multiply 57 x 42, we could do it by multiplying 59 x 4 for a first component. The second component would be the difference between the digits in 5 and 4, multiplied by the two in 42. The assembled components would then be 236/02/14= 2394. If we turn the problem around and went to multiply 57x24, we could do it by adding one to the five in 57 and multiplying by two to get 12. The next component would be the one that we left behind times the two in 24 + (5-2) times the four in 24. That would give us 2+12 equals 14 for second component.

The method lends itself to some shortcuts that are not available in the traditional vertical and crosswise system. There are some other nice features as well, one being that the method is subject to aliquot parts. An aliquot part is a proper divisor of an integer. For example, anything that you had to multiply where the units add up to five would count as one half when you took it over to the other side. For example, let's multiply 84x21. In this casem the units add up to five. Our first component would be 85x2 or 17. The second component would be 06, and the final component would be 04. The solution would become 17/06/04= 1764. We will see that aliquot parts become really important in the three digit world.

Notice that if we were multiplying 84x31 or 83x32, the units would add to five, and the difference between the numbers would be five too. Now the components can be assembled quite easily. The first component would be the vertical calculation of 8x3=24. The second component would be 5 x (the sum of the digits). In the one case 5x4=20, in the other 5x5=25. The final component would be the vertical calculation of the units' digits for each problem. In one case, we have 24/20/04=2604; in the second case, we would have 24/25/06=2656.

If the base is the same in each case, in either the tens' column or the units column, the multiplication is a simple two step process not much different than squaring. If you wanted to calculate 39x38, just multiply 47x3=141, and 72 to it to get 141/72=1482. By the same token, 93x83 could be done by adding 17 in front of the 3 in 93. Multiply that by the other 3 to get 519. Now link 72 in front of that 72/519= 7719. It is a two step process just like squaring is.

One other easy Aliquot part is when the digits add up to 15. The amount that you are carrying forward or back is 1.5 or 15. That works out nicely when multiplying. If you wanted to multiply 97x98, you could multiply 105x9 to get 945. Now link that with 8x7=56. Your final result would be 9506.

Remember the fact of bi-directionality also. In the case of 57x42, we could note that the tens' column adds up to nine. Now we carry that nine in front of the seven in 57, and double it. That would give us 194. To that we add in the difference of 5 in the units' column, multiplied by 4. The adjustment is 20, and 194 is changed to 394. The other link is 5x4=20. Our final result is still 2394.

The same rule would apply that you used when squaring. That would be to carry forward or backwards towards the smaller digit in the smaller number as much as you can. Unless the sum or difference from 10 is exactly the same on both columns ,don't try to get too clever with the cross products. Let's try some exercises, and see how it goes.

Exercises

Multiply these two digit numbers in your head. Just record your answers on paper as you go along.

1. 98x78
2. 87x98
3. 54x47
4. 65x57
5. 52x69
6. 74x86
7. 47x96
8. 67x49
9. 32x58
10. 21x91
11. 64x36
12. 23x94
13. 78x91
14. 48x72
15. 97x52

Answers

1. 98x78= 7644
2. 87x98= 8526
3. 54x47= 2538
4. 65x57= 3705
5. 52x69= 3588
6. 74x86= 6364
7. 47x96= 4512
8. 67x49= 3283
9. 32x58= 1856
10. 21x91= 1911
11. 64x36= 2304
12. 23x94= 2162
13. 78x91= 7098
14. 48x72= 3456
15. 97x52= 5044

4. Three Digits any Numbers

Three digit multiplications of numbers that are not squares is also possible using a sum and difference method. The easiest problem type after doing squares is numbers that contain partial squares. For example, if you had to multiply 315x215, the split would logically be 3/15 and 2/15. Clearly, you could do this as a fraction. The sum of this fraction is 5/15ths, which is the same as 1/3rd. If 15x15=225, the middle component is 1/3rd of that, or 75. Now just assemble 06/075/225= 67725. You could do the same problem in three steps with the sum and difference method. Add the 15s, and carry forward to the 3. Then multiply 33x2=66. Take the difference between 3 and 2, and multiply that by 15. Finally, square 15 to get the last component. The assembly would be 66/15/225= 67725. I don't know that one method really has an advantage in this case, but having an additional way to check your work that is independent of the first method is a huge advantage in terms of accuracy.

The secret about splitting the three digits so that you have a two digit part that adds up to 30,40,50, and so forth is just to find a combination where the units' or tens' column is comprised of tens' compliments. For example, if you were to multiply 852x728, you can immediately see that you have tens' compliments in the units' column. A closer look shows that the last two digits total up to 80. The math after that is quite simple. The components become 616/28/1456= 620256.

If the problem were changed so that the middle column had the tens' compliments you could solve it by making the split between the digits after the first two columns. For example, you could solve 528x287 easily by splitting it into 52/8 and 28/7. We know that 52x28=1456, or we can easily calculate that it is. Our components become 1456/28/5656=151536. The 28 comes from (8-7)x28. The 5656 comes from 52+28=80, added to the front of the 8 in 528, and multiplied by the 7 in 287.

You can also have aliquot parts with three digit multiplication. The parts that are especially useful are situations where the two digit column adds to 50, 75, 100, and 150. You always want to take advantage of a situation where the columns after you split them are 1, 5, 10, or 15 units apart, as they are all so easy to work with when you compute components. For example, 926x449. Split this into 9/26 and 4/49. Since 49+26=75, you will end up multiplying 9&3/4x4 to get 39 as your first component. The second component is 5x49, which is an easy aliquot multiplication. Half of 490 is 245 for your second component. All that is left is 26x49=1274. Once you have the components, the rest is 39/245/1274=415774.

If the numbers do not afford you an easy answer, you can still use the method. Just split your digits towards the smallest digit of the smaller number as a general rule.   For example, if you wanted to multiply 895x349, how would you proceed? I would split the numbers into 8/95 and 3/49. Then I would multiply 944x3 to get 2832 as the first component. The second component would be 5x49=245. Now we have 3077 to link with 95x49. That would be 3077/4655=312355. It looked worse than it was, as you could take it in smaller parts after you got your 3077. Just link 3077/(104x4)=31186. Now link 31186 with 45/45 or 495. That link would be 31186/495=312355. Break the numbers into small parts, and just keep building to get your answer.

Recall when we did two digit numbers that if the difference between the tens' digits was the same as the units' digit portion of the sum of the units digits, then our middle component became the sum of the digits in the smaller number multiplied by that difference. For example, 41x12 has a difference of three in the tens' column, and the units' digits add up to three. We could do this one straightaway as 4/(3x(1+2))/02= 492.

The same thing occurs in three digit multiplication. For example, let's multiply 724 x 226. We can see that the difference between the 7 and 2 is five. We can also see that 24+26 equals 50. The math after that is quite simple. Our components are 14/(20+26)x5))/624. That would give us a final answer of 14/230/624= 163,624. The middle component is 2(0) added to the number 26 to get 46, which we then multiply by 5 to get 230.

If we wanted to multiply 311x109, we could do it in a similar fashion. In this case, the difference is two, and the sum is 20. Now we just proceed to put the components together. That would be 03/38/99= 33,899. You can also use this in a aliquot parts fashion. For example, if we change the problem we just did, so that we were multiplying 211x109. In this case, the difference is one in the hundreds' column. The units' column adds up to twice that though. Now our components become 02/29/99= 22,999. The middle component became (2x10)+9. Note that it did not become what you would presume to be a logical one half of 38. If you play with this one a little bit, you will get a good feel for how it works, and it may be useful to extend the range of this little trick.

The same technique can be used in mixed digit multiplications such as 315x15. In this case, the 15s add up to 30 and 3-0= 3. From here we proceed. There is a first component, and it just happens to be 3x0=0. The second component becomes 3x(15)=45, or 30+15=45, to keep it in digit sum terms. To this, just link 225 to get 4725. We could have done this same thing with any combination that added up to 30 involving the sum of the last two digits of the multiplier and multiplicand.

For example, 314x16, 313x17, 312x18, and so forth would all have zero as the first component. The second component would only change slightly; it would be 3x16, 3x17, and 3x18 in that order. The last digit would be 14x16, 13x17, and 12x18 to complete the multiplications of these three numbers.

There are a lot of other combinations where this works. Three digits just exponentially increase the opportunities to apply this. Then there is the situation where the "units' digits" adds up to a multiple of 11, and the hundreds' digits are separated by that sum divided by 11. For example, let's look at 829x326. The split on this would be 8/29 and 3/26. It hardly lends itself to using fractions to find the components. Since 29+26=55, and 8-3=5, we can solve this much the same as we would a problem that summed to 50 and had a difference of 5 in the hundreds' column. Our components would become 24/59x5/29x26, with the 59 deriving from adding the 3 in 326 to each digit of the number 26. Our final result would be 24/295/754=270254.

Another easy situation occurs when the units' digits add up to 11 and the tens add up to the same number that the hundreds digits are separated by. The problem of 829x322 exemplifies this situation. The numbers 29+22= 51, and the hundreds' digits are 5 units apart. Can you guess what the cross product will be? The components of this one are 24/263/638= 266938. The 263 comes from adding 30 to the 22 to get 52. Now you multiply that by 5 to get 260. The three is tacked on after that, with no additional multiplications to it.

Let's do some exercises to increase our confidence and proficiency with three digit multiplications.

Exercises

1. 219x48
2. 999x819
3. 924x336
4. 518x302
5. 242x818
6. 511x219
7. 787x638
8. 126x319
9. 742x208
10. 123x63
11. 416x117
12. 915x616

Answers

1. (21/9)x(4/8)= 84/204/72= 10512
2. (99/9)x(81/9)= 8019/18x(81+09)/81=   818181
3. (9/24)x(3/36)= 27/396/864= 310464
4. (5/18)x(3/02)= 15/64/36= 156436
5. (2/42)x(8/18)= 16/372/756= 197956
6. (5/11)x(2/19)= 10/117/209= 111909
7. (7/87)x(6/38)= 4950/38/3306= 502106
8. (1/26)x(3/19)= 345/52/494= 40194
9. (7/42)x(2/08)= 14/140/336= 154336
10. (12/3)x(6/3)= 72/54/09= 7749
11. (416(x(117)= 04/84/272= 48672
12. (915)x(616)= 54/(228+6)/240=563640

5. Solve for Something Else

Sometimes, the best way to solve a math problem is to solve for something else. This is not an uncommon problem solving technique. It's done all the time in all sorts of applications not related to math. It is a great technique to learn and then use elsewhere as a lifelong problem solving skill. I can't think of a better subject to apply this thought process to than math. Let's go to a simple example that demonstrates the principle involved.

If we wanted to multiply 18x68, I hope your immediate response would be to think in terms of fractions. You have 1/8th and 6/8ths. The two fractions add to 7/8ths. You know your components to be 6/56/64. The answer is 1224. Another approach that you could use would be to say to yourself that with the addition of 1/8th, the fraction would become

8/8th, or the whole number one. The middle component would be the same as the last component, or 8x8=64. To get there, you could change 18 to 28, or multiply 1x8, and subtract that from 64 to get 56. In this case, we took an easy problem and made it more left handed. It does not always work out that way though.

If the problem had been to solve for 116x78, I would still think fraction. This time, if you added 1/16th to 1/16, you would make the second and third components equal at 8x16=128 each. This time, just adjust slightly, and subtract 8 from 128 to get 120 as your second component. So we solved initially for a second component of 128, and subtracted 8 to get 120.

The salient point is that the correct adjustment of one unit on one side of the split numbers was to adjust the diagonal number from it by the value of that number. In both these examples, that adjustment was 8, and it was subtractive. The adjustment would be additive if the addition of one more unit got us to a result that was almost instant. We could extend this out another increment just by doubling our adjustment too. All this should not be surprising, considering our experience with open carry methods.

Under that system, you manipulate your adjustments similarly. For example, if we wanted to find 38x21, we could take a similar approach. We know that if we had to find the cross product of 39x21, it would be 21 right under our nose as the units add up to 10, and the tens are one unit apart. What would it take to adjust that to get the cross product of 38x21? The adjustment needs to be subtractive. It should be the value of the tens' unit in the smaller number. So we adjust the 21 downward by 2 units. The cross product is 19.

If the question had been to solve for 47x32, we could easily do this one as 12/(32-3)/14=1504. If we were trying to solve for 48x34, the adjustment would be for 2 units from 34 and would be additive. The components would become 12/(6+34)/32= 1632. We have applied the occasion to solve a problem by solving for something else to fractional situations and to open carry situations. Where else would they be applicable?

In a previous book, we covered numbers that were close to a base of one, 10, 100, and even 1000. I hope you've asked yourself the question, "What would the adjustment be if we were working in that system?" For example, if we wanted to multiply 115x115, we could do it by noting that each number is 15 units over a base of 100. Then we would add 15 to one of 115s. That would increase it to 130. To that, we link the product of 15x15. The final result would be 130/225= 13,225. What if we were presented with the problem of finding the product of 215x115? When you're first presented with this problem, you might look at it and think to yourself that if only that 2 in 215 was a 1, you would have this one knocked out in no time. The truth is that you can do that by taking the prefix of 115x115, the 130 and adding 115 to it to get 245. Now link in the 225 to get 24725. If the question had been to solve for 315x115, you could do it in a similar manner. This time, just add 115x2=230 to 130 to get 36/225=36225. In this case, it is easier to get this answer than it is to solve for 215x115.

Remember that the last two digits don't have to be identical; they were in this example only. You could solve for 112x213 by adding 112 to 125 to get 237. Then you would link the product of 12 and 13, which is 156. The final result would be 137/156= 23,856. 313x112 would result in 35056.

If you were looking at multiplying 313x412, you could start off with the prefix for 313x312. That would be 325x3= 972. To that, add 313 to get 1288/156=128956. What we really did with all these little fixes is come up with a way to get the easy answer, and then adjust our way easily into an answer for a different problem. We have applied this method to problems involving fractions, problems involving the identity property, and problems that are close to being an open carry situation. There is yet another way, to which we can apply the same principle.

The method in the case of base related problems is especially powerful. You can apply it to a mixed digit calculation such as 107x54. The answer would become (5x54)+(5x61)/28=270+305/28=5778. The 5x54 is the difference between 10&5 multiplied by 54. The adjustment is 5x(54+7)=305. The 28 is from multiplying 7x4. 324x54 would work out to be:
(32.5)x54=1458+(58x5)=1748/16=17496. The 58 would be the result of adding 54&4, and the multiplier 5 is from 54 having a base of 5(0).

Another simple example of solving for one thing to get the answer to something else comes along in the form of problems that are close to the shortcuts we are all familiar with. For example, we had to solve the problem of 16x15. We could first solve for 15x15. You will recall that this is a real easy multiplication. The answer is 225. What do we have to do to correct 225 to get 16x15? All we have to do, it turns out, is to add another 15; and that is considerably easier than doing the actual multiplication. Could we do 14 x 15 in a similar fashion? The answer is yes, in this case we would square 15, and subtract 15. The result would look like this: 225-15=210.

You could also apply this incremental approach to multiplications involving reciprocals. For example, 27x72 involves multiplying a number by its reciprocal. With reciprocals, the first coefficient and the last coefficient are identical. In this case, they're both 14. The middle coefficient is always going to be the square of the two digits added together. The middle coefficient in this case is (2x2)+(7x7)= 53. If you multiply 27x73, hopefully you would observe that this is within one number of being a reciprocal. One way that you could solve it would be to take 1944, and add 27 to it. That would give you 1971. Even better would be to take that additive change of 1 and add 1x2 to it. In this case, the adjustment is again the amount of change multiplied by its diagonal opposite, which is 2.

There are other ways that you could use this incremental approach. We know that the number 11 has some unique properties. For one thing, the sum of the odd coefficients equals the middle coefficient any time you multiply a number by 11 or a multiple of 11. You may occasionally look at a problem and notice that you are almost multiplying by a multiple of 11. Let's take the example of 23x64. That 23 is one unit away from being a multiple of 11. Let's go ahead and multiply 64 x 23 and treat the middle component for a moment as though it were the middle component of 64x22. We start out with a first component of 6x2=12. This calculation is the same for either problem. The second component of 64x22 is real easy to get. It is 2x(6+4)=20. It's just a single multiplication of the sum of the digits in 64. We know that the cross product of 64x23= ((6x3)+(4x2))=26. The correction to the 20 has to be to add 6 units. You get that by multiplying the change of one unit by its diagonal opposite, which is 6. The final component is going to be 4x3=12, instead of the 8 you would get from 2x4=8 in 64x22. Our final answer on this one would be 12/26/12=1472.

So now we have a whole new method based on the unique properties of 11 and multiples of it. You could use this one anytime you see numbers that are close to being a multiple of 11. You can use it in a row situation, or you can use it in a column situation. It is also bidirectional, so it has enormous potential to be a very effective tool. The method would be especially effective when the adjustment is only one unit because your correction then is always a single digit. Just for fun, let's look at a problem like 98x97. We already know that we can solve this using a under a base method. 98 is two units under 100, and 97 is three units under 100. Now we subtract three from 98, or subtract two from 97, and either way, we get 95. To this, we tack on 3x2=06. The easy final answer to this is 9506. We could check our work by multiplying 9x9 to get 81, and then we could link it with (16x9)-9=135. Now we have to put together 81/135/56. That gives us 9506. This approach was not too much more difficult than the first method we used to solve a problem.

I deliberately selected this problem to make another point. If you think about it for a moment, you already had multiple of 11 in the first column. In this case, you could have written your components straight across as 81/9x(8+7)/56= 81/135/56= 9506.

I am trying to make the point that using a multiple of 11 in a column is quite easy, since you multiply your components in a column fashion as well. If you don't look at the problem in both a row or column format, you may miss the obvious. The further point is that having the same base in either the tens' column or the units' column presents the problem as a multiple of 11 situation. Sometimes we get so caught up in the narrative of having the same base that we miss other interpretations that can be quite insightful. In this situation though, remember that the first and last coefficients will not add up to the middle coefficient, and the horizontal calculations will add up to the cross product.

For practice in this section, make up some problems that are "almost" situations. Then correct for the increment to get the right answer. Keep playing with the numbers. That is how you will get better, more confident, and more comfortable with them.

ABOUT THE AUTHOR

John graduated from the U.S. Merchant Marine Academy, and has an MBA from the University of Minnesota. Currently he lives in Apple Valley Minnesota. As a Merchant Mariner he traveled all over the world. His web site about vedic math which you may find interesting is at www.binomialblvd.com

John ran into a Russian merchant mariner in Istanbul many years ago. They talked shop and discovered that they used some of the same books to learn their trade. When they looked at the Calculus books each had used in college the comparison fell apart. The Russian had a trim little 150 page Calculus book. John had a 600 page monster book but knew far less Calc than his new friend from Odessa. Ever since then John has had a phobia about 600 page math books, that is why his are so short.

One good thing that happened from this encounter was the Russian gave him a mental math book by Trachtenburg who ironically was originally in charge of the czar's shipyards in Odessa. From there John got the mental math bug, and along the way came to believe that math phobias are the product of the way we teach the subject. From there he stumbled upon Vedic Mathematics and was so fascinated by it that he started to teach vedic math for kids in his spare time. He eventually wrote a series of four books on the subject.

John Carlin

www.ingramcontent.com/pod-product-compliance
Lightning Source LLC
Chambersburg PA
CBHW072045190526
45165CB00018B/1736